NOT THE SOLUTION

CHAD M. ROBICHAUX &
JEREMY M. STALNECKER

Three Clicks Publishing

Not The Solution

© 2023 All Rights Reserved

Chad M. Robichaux and Jeremy M. Stalnecker

Limit of Liability/Disclaimer of Warranty: While the publisher and authors have used their best efforts in preparing this book, they make no representations or warranties with respect to the accuracy and completeness of the contents of this book and specifically disclaim any implied warranties of merchantability or fitness for a particular purpose. No warranty may be created or extended by sales representatives or written sales materials. The advice and strategies contained herein may not be suitable for your situation. The author and publisher are not engaged in rendering professional, legal or medical services, and you should consult a professional where appropriate. The authors and publisher shall not be liable for any loss of profit, nor any personal or commercial damages, including but not limited to special, incidental, consequential, or other damages.

All Scriptures taken from the Holy Bible: New International Version. NIV. Copyright © 1973, 1978, 1984, International Bible Society. Used by permission of Zondervan Publishing House. All rights reserved.

All rights reserved. No part of this book may be reproduced or transmitted in any form or by any means except as permitted under Section 107 or 108 of the 1976 United States Copyright Act and with written permission from the author. All materials are legal property of Chad M. Robichaux and Jeremy M. Stalnecker. Unauthorized duplication is strictly forbidden and punishable to the maximum extent under applicable law.

ISBN: 979-8-9882791-9-8

Edited by: Nicki Thompson

Published by: **Three Clicks Publishing**

Learn more information about the authors at:

www.PathToResiliency.com

Contents

Introduction 5

It Happens in the Dark 15

The Mighty Oaks Story 25

The Real Cost of Suicide 33

The Big Picture-A Path Forward 41

What Now 55

Hope 57

Introduction

By Chad Robichaux

Founder, Mighty Oaks Foundation

I'm desperate. Depressed. Hopeless. I've lost my sense of purpose. What is there for me to do now? What am I even fighting for anymore? I'm no longer productive or effective. I've become a liability. A burden. A frustration. A waste of space. I'm just dragging everyone down around me. Everyone would probably be better off without me around to make everything harder. I should just remove myself from the equation entirely so that I won't be spreading around my pain, anguish, and negativity. With me gone, my family will be happier. My wife and kids will live fuller lives without me there to ruin everything. My friends will have more fun.

I can't tell you how many times I've heard words exactly like this. To be honest, I even felt this way myself once. "Maybe my family will be sad without me, but they'll be better off." How tragic that this hopeless thought has found a home in the hearts of over 20 veterans a day.

But by the grace of God, I realized that this is not reality. Suicide is not the answer. And have now committed my life to helping others realize this as well.

You don't make the pain and anguish go away by choosing to remove yourself from the equation through suicide. In fact, the exact opposite is true. All you do is multiply the pain and anguish, spreading it around and transferring it to those closest to you.

You transfer it to your spouse, who finds themselves relegated to a life of guilt, confusion, and heartache.

You transfer it to your children, who mourn years of lost time and opportunities, constantly thinking about how things might have been or what they're missing out on.

You transfer it to your friends, who wonder what they must have done for you not to trust them to support you.

You transfer it to your community, who would have benefited from your gifts and talents, but won't ever get that chance.

You transfer it to future generations, who see your example and decide that if suicide was a way out for you, then maybe it's a way out for them as well, starting a chain reaction of devastating decisions destroying innumerable lives and legacies.

Introduction

The pain and anguish don't go away just because you do. Instead, they grow exponentially and spread to everyone around you. The exact opposite of what you thought you were doing.

As on many Sundays over the past decades, I was sharing the story of my own struggle at a church in Oklahoma and ended by offering a solution through our programs at Mighty Oaks to anyone who may be struggling from the military and first responder communities, to include their spouses.

In attendance were Heather and her husband Pete, a Marine combat veteran diagnosed with PTSD and dealing with a devastating lack of purpose and hope. But like many Marines, he thought, 'Why would I fill a spot at Mighty Oaks and take that seat away from someone who surely needs it more?' So, Pete did not accept my invitation to get help and find a new mission, nor did he heed the urging of his wife, Heather.

I'll never forget the call I got six months later from the Pastor Ron Woods of The Assembly Church, telling me what had happened to Pete. In a dark moment, Pete stood in the back of a pickup truck surrounded by police. As he held a pistol to his head, his final words were, "Tell my wife I love her and that I'm doing this for her."

Then he pulled the trigger.

Pastor Ron requested that my wife and I come to speak to Heather in the aftermath of Pete's suicide and when we met her, she expressed to us a feeling of deep obligation to speak to other struggling veterans about what Pete had believed in those final moments—that he was doing this for her.

For years, both in person and through video, Heather has bravely sat in front of thousands of warriors to explain how backwards it is to think your suicide will improve the lives of your loved ones.

She talks about holding his lifeless body when he laid on the metal table of the morgue and putting her head on his chest trying to hear a heartbeat. She talked about calling his mom, how she screamed, and how neither of them will ever be the same. Heather was injured at the time and when she needed him at home, Pete had been there. But now he had left her, devastated, to pick up all the pieces. Pete thought that he was ending the pain but all he was did was multiply and transfer it to all the people who loved him.

To this day, even some of the most hardened combat veterans and first responders break down in tears when they hear Heather's story because they are hearing her speak directly to the misguided thoughts they are having. I've seen several literally run out of the room because they can't stand to hear her words.

I believed that same lie once, and hearing this story from Heather and seeing thousands come through Mighty Oaks over the years, all believing the same lie, makes me ask how many others the devil has whispered this lie to, and have they believed it—"Maybe they will be sad without me but they will be better off." This is a lie straight from the pit of hell to destroy you and everyone you love. Had I taken my own life as Pete did, there would be no Mighty Oaks Foundation for thousands who have attended the recovery programs. I would never have spoken to the half a million active duty troops on bases, written numerous bestselling books that share principles to help others, or donated three-hundred thousand copies to those in need. I wouldn't have spoken to millions of people in churches and broadcasts to share stories of hope with others who struggle as I did. None of that would have happened.

I would have left a wife and three children, who would have learned that suicide is how you solve a rock-bottom problem. And my dear wife Kathy would have been left to pick up and hold all the shattered pieces for years of a life and family forever changed.

But I chose a different way.

Now, we've been married 28 years, we have three married kids, grandkids, and even a brand-new baby girl that we recently adopted. I relish watching our kids and their families find success and serve in their communities.

I can't imagine where they might be if I had left them, misguidedly thinking they would be better off without me. I'm glad I was willing to set an example of digging in and fighting when things got hard rather than quitting and leaving it for others to deal with. And I'm grateful to be able to prevent a little bit of this tragic cycle with hundreds of thousands of troops and first responders who are also struggling.

I'm here to tell you that suicide is not the answer.

Instead of surrendering to the purposelessness, hopelessness, pain, and anguish, thereby spreading it around to everyone in your life, you can find a new purpose. A new hope. A new mission to give your life meaning. You have so much still to give, you just don't yet know how to give it.

But God knows. God sees how much you can give and who still needs you, and He's just waiting for you to be ready for your new mission.

On the same weekend that Pete said no to my invitation to find a new mission and purpose, Reed Hastey said yes. He sat in the same congregation as Pete, but Reed chose a different path. An Army combat veteran with debilitating anxiety, PTSD, and perpetual suicidal ideation, Reed spent many weekends being held in VA inpatient clinics. He felt humiliated to have gone from a respected, effective, and influential leader of essential combat teams to being locked up in an inpatient clinic as a danger to himself and a man incapable of even the most basic functions for life, like being trusted with his own shoelaces.

After a significant inner battle, he managed to come hear me speak at Pastor Ron's church and while there, Reed saw a glimpse of hope through what God had done for me. He gave me and Mighty Oaks a chance, although because of his anxiety it took him nearly six months to be able to get on a plane to fly to a program. But while there, Reed surrendered his pain, his struggles, and his life to Jesus and everything for him radically changed from that moment on. He immediately came back through to Mighty Oaks again, this time for our leadership program to become a Mighty Oaks Team Leader. After overcoming so many challenges, Reed had found his new mission as one of the most effective team leaders at the Mighty Oaks Foundation, even becoming the head trainer for new Team Leaders.

In subsequent years, Reed has blessed his community by establishing 13 Mighty Oaks Outposts, serving over 500 military and first responders who meet weekly for support. He is directly responsible for more than 130 Tulsa police officers find their own renewed sense of hope and purpose through the Mighty Oaks Foundation programs within a two-year period. The Tulsa Chief of Police said this has changed the culture of the police department and as a result, has impacted his entire city.

Reed was once a man so paralyzed by anxiety and depression that he struggled to even live, and took six months to simply travel to California for our program. But Reed recently traveled with me to Ukraine, flying across the world to Poland and driving 15 hours across the Ukraine border and east into the Russian occupied Red Zone to share his testimony with Ukrainian soldiers to inspire hope and purpose amid anguish and desperation. It was a powerful moment for me to see Reed's journey come full circle as we sat in an old building, hidden in a muddy forest, with hundreds of Ukrainian soldiers gathered around him in a circle as he began speaking to them. Then and throughout the trip, indirect fire from the Russians impacted in the distance. You could not only hear it but feel the ground shake, with closer impacts being felt throughout your body. All the while, Reed was cool as a cucumber knowing it was his mission to help these people find hope in the midst of their suffering. And he had his shoelaces on the whole time.

In the military, we are trained in certain tasks for use on certain missions. Just because the mission changes doesn't mean our skills can't be used or that our experience isn't valuable. It just means we might need some training for our new mission. And as with all the other training for all the other missions, the ones who have gone through it all before are ready to help you every step of the way to prepare you for your new mission.

Then one day, you'll be the one who went through it all and came out on the other side stronger and more capable, and part of your new mission will be helping others discover their new mission rather than surrendering to deceptive and destructive thoughts.

Saddam Hussein once predicted that while we might win the war, our soldiers would die by their own hands. Let's not make a murderous dictator right, and press on to the battle ahead.

It Happens in the Dark

A Biblical Look at Depression, Despair and Deliverance

On March 19, 2003 I had the privilege of moving from Kuwait into Iraq in what would become known as "Operation Iraqi Freedom" or OIF 1. This was a moment that we had been planning for during the previous year and had aggressively war-gamed throughout the months of February and early March. We spent so much time preparing, in fact, that nearly 20 years later I can still remember the operational terms and call-signs along with most of the visual markers that we would use during our movement. Each step, the placement of each vehicle, and the exact timeline were memorized from the top of the command chain down to the bottom. Nothing was left to chance. The plan was to begin our movement as the sun started to rise in the early hours of March 19th and, if you had asked me on the 18th, I would have told you that this was going to be a simple, daylight movement with little chance for anything to go wrong.

I have always been a little naïve. After months of planning to move with the sun in the sky, we instead received the order to go during the darkest part of the night! Nothing we had planned-except moving into another country-applied and the mission was to just "get there." While I am sure there were strategic reasons for this last-minute change, one thing became very clear: When you find yourself in the dark, even when you start with a good plan, very little makes sense and even "getting there" becomes overwhelming.

As I have gotten older and lived through some battles that did not include invading other countries, I've discovered that fighting in the dark is ALWAYS hard even when the battles are emotional, or spiritual, or relational, instead of physical. I believe that most of us start off with a good plan but that the events of life and the actions of others can push us into a darkness that we didn't see coming. So many things stop making sense and doing anything other than just "holding on" feels impossible. It is often in these dark, unexpected times that we make permanent decisions that will destroy our future and greatly impact the future of those closest to us. The dark makes everything worse and there is nothing darker than an overwhelmed soul.

What I find interesting about this is that everyone experiences the dark but not everyone allows it to overwhelm them. I think one of the lies we tell ourselves when life becomes uncertain is that no one can possibly relate. We believe that what we are living through is unlike anything anyone else has experienced and therefore gives us permission to do things we know, at least in our rational moments, are not right. These thoughts then lead us to the conclusion we are probably the ones at fault for our situation and the best thing we could do is simply remove the problem-US! We rationalize that normal people don't have these kinds of problems so we must be broken. We conclude that the world will be a better place if we are no longer in it. It is this kind of thinking that causes many to reject the Bible or a relationship with God as a solution because they have come to believe that God, the Bible, and the Church, are only for people who "have it all together." It doesn't apply to us though.

In moments like this I like to share one of my favorite stories from the life of Jesus.

The story can be found in the Bible in Mark 14:34-36. I will paraphrase the story here but encourage you to take some time to read the whole thing. It is an incredible narrative and so clearly illustrates the humanity of Jesus Christ. The story takes place on the night that Jesus would be arrested in a place called Gethsemane. This was a place He would go alone to pray and on this night, he brought his disciples. We could spend a lot of time talking about the disciples but suffice it to say that the only thing Jesus asked them to do was stay awake, and the only thing they did was fall asleep! As He was approaching the crucifixion, it seems that Jesus wanted to have people around him that cared about what He was going through. Instead, He ended up very alone and overwhelmed by what was about to happen.

If you are familiar with this story, you know that Jesus used this time to pray to God, the Father. Again, there is much that could be said here, but what we must not forget is that Jesus is God the Son and one with God the Father. At this moment in time, though, there would be a temporary separation between the two as Jesus took sin on Himself to make it possible for humanity to have a relationship with the Father.

The takeaway for us is this:

Jesus is God and, as such, is perfect.

Jesus was experiencing rejection, separation, and anguish because He was getting ready to engage in the most important act in history-the redemption of mankind.

In spite of who He is, the darkness of what He was about to do began to overwhelm Him.

The anguish of this moment was so great that Luke 22:44 tells us that as Jesus cried out to God the Father He sweat "great drops of blood." As all of this was happening Jesus must have remembered the prophecy of Isaiah 53:3 that said He would be, "a man of sorrows and acquainted with grief." And yet through everything, He persevered and accomplished the mission that brought Him to earth.

What I love about this story is just how relatable it makes Jesus to my life! He is God, did no wrong, was giving everything to serve others, and STILL experienced sorrow and grief. If Jesus experienced these things, then we must believe they will be a part of the normal life experience. We don't want to live through any of this, but when it comes, we can be comforted knowing that we are not alone and that there is a path forward.

So, what lessons can we take from the story of Jesus as He lived through the darkness of rejection and anguish?

> 1. We are not alone. If even Jesus had to endure these things, then there is not a person on earth who will not. You are not broken because you are dealing with something hard (whether you created that hard thing or not) you are simply human. Don't ever believe the lie that there is no one in the world who understands.

> 2. The mission is what carries you forward. It is during this time that the Bible records Jesus saying to God the Father, "Not my will but thine be done" (Mark 14:36). Jesus had

previously declared that His mission was to pay the price for sin on the cross (Luke 19:10) and it was this mission that carried Him forward when it would have been easier, and made more sense, to quit.

3. None of this was about Him. Since the mission was clear, Jesus lived for the benefit of those He came to redeem. We move through the darkness because there are people in our lives who need us to do it for them. Jesus, as God, did what He did for us. We need to live our lives and make decisions for others (Philippians 2).

4. Timing does not always make sense. That's ok though because it's not about you! Jesus clearly understood the timeline that He was operating on, but no one else did. That's why the disciples fell asleep. They didn't realize they had anything to worry about. Why do things happen when they do, and why do they last as long as they do? Only God knows the answers to either of those questions so our focus should move from "why" to "what now?"

5. You must look outside of yourself for Hope. As Jesus prayed, He was declaring that His source of hope and direction was the Father. Faith in one who does not change and

has it all figured out is the only way to move forward during the dark, anguish of soul times of life.

Some truths to remember:

Psalm 34:18 The LORD is near unto them that are of a broken heart; and saves those who are crushed in spirit.

2 Corinthians 12:9-10 And he said unto me, My grace is sufficient for thee: for my strength is made perfect in weakness. Most gladly therefore will I rather glory in my infirmities, that the power of Christ may rest upon me. 10 Therefore I take pleasure in infirmities, in reproaches, in necessities, in persecutions, in distresses for Christ's sake: for when I am weak, then am I strong.

1Corinthians 10:13 No temptation has overtaken you except such as is common to man; but God is faithful, who will not allow you to be tempted beyond what you are able, but with the temptation will also make the way of escape, that you may be able to bear it.

Romans 8:31 What then shall we say to these things? If God is for us, who can be against us? 35 Who shall separate us from the love of Christ? Shall tribulation, or distress, or persecution, or famine, or nakedness, or peril, or sword? 37 Yet in all these things we are more than conquerors through Him who loved us. 38 For I am persuaded that neither death nor life, nor angels nor principalities nor powers, nor things present nor things to come, 39 nor height nor depth, nor any other created thing, shall be able to separate us from the love of God which is in Christ Jesus our Lord. God loves you and has made every provision for you in Christ.

On that dark night moving into Iraq in 2003, our battalion successfully accomplished our mission despite unforeseen obstacles (like changing the time of attack) and often overwhelming frustrations. We could not see, were exhausted, and had never done anything quite like this before. But we still came out on top. Why?

We did not have a choice! Quitting was not an option anyone offered.

We had already decided we weren't going to lose two years before getting there as we trained in Southern CA. The decision to move forward despite what might happen had already been made.

We trusted those who had a bigger picture of the battlefield. They knew what they were doing.

War and life are not that different. You may find yourself in the dark, but you don't have to stay there.

The Mighty Oaks Story

"To Restore the brokenhearted through Christ, to build leaders of leaders to rise up from the ashes; they will be called Mighty oaks of Righteousness."

-Isaiah 61

It is always interesting to see what people do when they find the hope for which they have prayed. Many will view this hope as a gift given to them alone and use it to move forward in a way once thought impossible. Others will acknowledge that hope exists but will not allow it to take hold, deciding instead to return to a life of desperation simply because it is familiar and, in an odd way, easier. There are some though who understand that the gift of hope is not given to be owned by an individual but that it should be appreciated, used, and then handed to someone else who is need of a miracle. One thing that does not change is the impact that one's response to hope-hold it, ignore it, or give it away-has on the world around them. Many people have asked why The Mighty Oaks Foundation has had so much success helping the broken find restoration. The answer is simple: from the very beginning our philosophy has been that no one owns hope and that, once found, it should be given away to as many people as possible.

As Chad Robichaux and his family began to experience the restoration that a relationship with and hope in Christ will provide, they started to look around for other individuals or organizations who were helping families like theirs. While there was no shortage of help for just about every hurting group of people, the help to Veteran and military families was extremely hard to find. Even the churches that had military focused ministry did not deal with the issues of trauma and post-traumatic stress caused by combat deployments. When he tells his story, Chad often says that when he could not find anyone else serving the Veteran community from a position of faith, he stepped back and asked the question, "Why not me?" This question along with a conviction that God was leading is what started the process of developing what would become the Mighty Oaks Foundation. It was during this time that both Chad and Kathy on separate occasions came across the passage of scripture that would confirm this decision and become the scriptural model for the Mighty Oaks methodology:

> *"To Restore the brokenhearted through Christ, to build leaders of leaders to rise up from the ashes; they will be called Mighty oaks of Righteousness." -Isaiah 61*

We are several years past the founding of Mighty Oaks but the process that God used to bring it about and the verses that He provided to express the vision have not changed. Mighty Oaks is a pay-it-forward movement dedicated to seeing restoration take place in the lives of those who have served our country. The mission statement of Mighty Oaks is:

> *"To serve and restore our nation's Warriors and families, who have endured hardship through their service to America, and to help them find a new life purpose through hope in Christ and our resiliency and peer-to-peer recovery programs."*

From one family deciding to pay forward what God had done in their lives we have seen more than 4,000 men and women attend an in-person Mighty Oaks Session serving both Veterans and their spouses as well as active-duty military members who are attending our programs on official orders. Additionally, we have distributed more than 100,000 books and spoken to over 150,000 active-duty military members and Veterans at events across the country and have become a leading voice in Veterans Advocacy at the highest level of government.

Now, I would imagine you are reading this book for one of three reasons:

1. You are struggling with thoughts of suicide or have attempted suicide and want to move forward in your life.

2. You have a friend or loved one who has struggled, and you are trying to figure out how to help them.

3. You are involved in a local church ministry and need to know what to say to those in your church struggling with suicidal thoughts or attempts.

And, if you fall into one of those three categories you are probably wondering why we would take the time to tell our origin story. It's an interesting story, but there is a more important reason that it's included in a book about suicide. We will get to the methodology that we have used to help thousands of struggling people a little bit later in the book, but we tell the story of our beginning to illustrate what's possible! The interesting thing about Mighty Oaks is that it was not started by professionals. It was started by a family and then by others that joined them, who understood that they did not need to remain broken and without hope. The thousands of people who have been helped through the Foundation have been helped one person at a time as those who have moved forward in the face of trauma, broken relationships, and thoughts and attempts of suicide have told others struggling with the same how they too could move forward. It is not complicated, but it does require a decision to listen, accept help. This decision should lead to the realization that, in accepting help, you now have a responsibility to help the person coming up behind you. It is this philosophy-the learn, grow and pay-it-forward philosophy-that allows those who find themselves in the darkness of depression and loss to stand up and live a life of purpose they never thought possible.

Perhaps you are still not convinced and would say something like: "Well that is all well and good, but my situation is different. I am in a really bad place and there is no way that the others you have helped could relate to what I am going through." We will do our best to communicate in the pages that follow that you are not alone, but let's end this chapter with a statistic, an anecdote, and a story that will hopefully encourage you to keep reading and apply the lessons that will be taught.

First, the statistic. For those who are not aware, the Veteran suicide rate in the United States is approximately 20 per day.[1] The Active-Duty suicide rate is between 1.5 and 3.5 per day depending on the methodology used to quantify suicide data.[2] These numbers mean that those who have served in the military are 22% more likely to take their lives than the general population. This is the group we work with each day. This statistic, along with many others, has driven us to develop a methodology that can, if followed, bring hope and healing to the most suicide at risk group of people in our country.

Next, an anecdote. Anecdotes are interesting because they are unprovable stories told to support facts. So, anecdotally, many of the 4,000 students who have attended a Mighty Oaks program started their week by saying they would take their life at the end of the week if they did not find something that would give them hope. Hundreds of times students have shared with us that they only attended a program to prove that it would not work and had already decided what they would do to end their lives when they got home. These stories are told through tears and end with a statement like, "Now I have hope, I want to live and believe that I can and am excited about life for the first time in a long time." Many of these stories can be found in video form on our website or the Mighty Oaks Warrior Programs YouTube channel. The lives we have seen changed have struggled with the horrors of war, abuse as children, and substance abuse. Yet in spite of it all they have received a hope that they are excited to pay forward for others.

Finally, a story. I met Dan on his first day as a student at one of our Programs. I didn't know much about him at the time except that he was a retired Marine Sergeant Major. As the session began, Dan seemed just like the other men in attendance. He had served proudly and honorably in the Marine Corps but now looked like someone without hope who was carrying the weight of nearly thirty years of service and a lot of brokenness on his shoulders. It was clear he didn't want to be there, but also that he had nowhere else to go.

As the week unfolded and he became more comfortable with us and the process we were taking him through, Dan shared the details of his life's story and what brought him to his current station in life. For nearly thirty years, Dan served with honor and distinction in the Marine Corps, achieving the highest enlisted rank possible: the rank of Sergeant Major. He served in both peacetime and war, leading thousands of Marines in locations around the world. Dan is a humble guy and shared this story as just a series of facts. The rest of the story was told with a great deal of emotion.

Even though he had a career that spanned three decades and was successful from nearly every perspective, it did not end that way. Broken relationships and a change in job responsibilities caused him to lose his way and begin abusing alcohol to the point where retirement became a mandatory event. He left the only thing he had ever really known and was now without purpose or direction. It wasn't long before life had so little purpose that the only thing that made sense was taking his life. It was a suicide attempt that led him to us. A failed attempt caused his family to reach out to Mighty Oaks, and his first stop after recovering from a month in the hospital was one of our programs. Dan is doing great now and is working to rebuild his life and move forward in a meaningful way. Before that was possible though, he had to come to a place where he could learn the path forward from others who had walked this ground before him. He decided that he would stop being a victim of his own bad decisions and now works to pay-it-forward for others.

So, in case you missed it, here is the reason for a chapter on the founding of a Veterans organization in a book on suicide: there is HOPE if only you will DECIDE to move forward. That is our story and that can be yours!

We will get to the way forward, but first let's take a look at the problem.

The Real Cost of Suicide

"My family will be better off without me."

"No one will miss me when I'm gone."

"I can never get over this. I have made too many bad decisions to keep going."

"I just want the pain to stop."

Since you are reading a book about suicide, convincing you that suicide is a problem is probably unnecessary. You or someone you care for may have even made one of the statements above, believing sincerely that it is true. What is so often missed when we have these discussions though, is the impact beyond the death of the one who decides to take their life. Only when we get the full picture of what is really at stake can we intervene in a meaningful way (if we are trying to help) and make better decisions if we are struggling.

Here are a few things to consider:

The Problem is Getting Worse

Recently, the Veterans Administration established a task force to end suicide called, "The Presidents Roadmap to Empower Veterans and End a National Tragedy of Suicide (PREVENTS)." In the introduction to their published plan the, severity and growth of the problem is given:

"The annual number of deaths by suicide in the U.S. increased by 35% from 1999 to 2018, with an estimated 48,344 deaths in 2018. Suicide is currently the second-leading cause of death for people ages 10-34 and fourth for ages 35-54. The number of deaths by suicide among Veteran and service member populations is equally concerning. In recent years, more than 6,000 Veterans, Guardsmen, active-duty Service members, and Reservists died by suicide each year-more than were killed in action in the Iraq and Afghanistan conflicts from 2001 to 2014 combined. The overall Veteran suicide rate is 1.5 times higher and the female Veteran suicide rate is 2.2 times higher than the general population's suicide rate after adjusting for age and sex. While shocking, these numbers pale in comparison to the number of individuals who attempted suicide: an estimated 1.4 million in 2017 alone."[3]

While these numbers should be startling to anyone really paying attention, what they do not account for are the families, communities, and workplaces that are also, at an increasing rate, dealing with the decision to commit suicide. We get a better understanding of the broader impact of suicide when we realize that, in the United States alone, the economic impact of suicide each year is roughly $56.8 billion![4] Clearly lives are more important than dollars, but understanding the economic impact helps to quantify just how deeply felt, in all corners of society, is that loss of life.

The Impact on Those Left Behind is Severe

A common statement made by those contemplating suicide is, "My family will be better off without me." The belief seems to be that taking ones life ends the pain they are experiencing while simultaneously putting everyone else in their world in a better position. There may be thoughts of relationship conflict, financial issues, embarrassment because of bad decisions, addiction, or some other life issue, followed by the conclusion that, "everything would be better if I would just go away."

During each of the Mighty Oaks sessions we play the video testimony of a wife who lost her husband to suicide. She tells the story of their marital ups and downs which included many of the issues listed above. As she talks about the day her husband died, describing even the moment she had to identify his body, she makes the following statement:

"I would do it all again, all of the struggles and challenges, just to have him back. My husband thought that he was ending his pain and leaving me in a better position. When I had to call his parents and tell them what happened, it was in the moment that I realized SUICIDE DOES NOT END YOUR PAIN-IT SIMPLY TRANSFERS IT TO THE ONES YOU LOVE."

Since statistics sometimes help to make things "more real," here are a few to ponder:

1. Children and teens who lose a parent to suicide are three times more likely to commit suicide than children and teenagers with living parents.[5]

2. Adults who lose a friend or family member to suicide are at a higher risk than the general population of attempting suicide.[6]

3. Those who lost parents to suicide are nearly twice as likely to be hospitalized for depression as those with living parents.[7]

4. Losing a parent increases the risk of a child committing a violent crime.[8]

5. Families experience high levels of rejection, shame, and stigma.[9]

6. Suicide-bereaved families experience intense guilt or feelings of responsibility for the death.[10]

7. One-quarter of people bereaved by suicide experience elevated levels of depression and stress, and one-fifth have elevated levels of anxiety.[11]

8. Psychosomatic reactions included physical or severe abdominal pain, loss of appetite, low energy levels, and sleep disruptions[12]

This list of researched consequences to the "ones left behind," could go on and on. What we see from this small sampling is that in truth, pain does not end when someone takes their life. Instead, it is passed on to the ones they say they love.

We are leaving the ones we say we love alone to pick up the pieces.

As we close this chapter, I want to appeal specifically to the one contemplating suicide. Maybe you have some of the things this chapter opened with and, right now, you are saying them sincerely. You believe that by removing yourself, the pain will end and everyone else will be better. There is hope and a path forward in the rest of this book, but none of it matters if you don't decide right now that you are going to get help, figure it out, get better, and continue to make a difference in the world. Even if you do not feel like any of those things are possible right now, at the very least I would ask you to understand that the "reasons" you are giving yourself for ending your life, even though they feel true, are not true at all. The only thing that ending your life will do is transfer your pain to those left behind-those left to spend the rest of their lives trying to figure out how to pick up the pieces to your permanent solution. Whatever your reasons right now, they are temporary. The decision to end your life is forever.

Twenty-four years after the death of his daughter by suicide, Art Linklettter, the radio and television personality, made this statement:

> "The word 'suicide' is without a doubt one of the most dreadful expressions in the English language. People wince at the sound of it and avoid using it to describe the tragic death it implies. Leprosy and cancer are spoken of in the same hushed tones. And yet it must be faced squarely and discussed openly because it has become one of the leading causes of death among both the young and the very old in this country. No one can be

sure of the exact figures because suicides are deliberately misreported and misdiagnosed as an accident. *My own personal experience with it is still a nightmare. The death of my nineteen-year-old daughter, Diane, after experimenting with LSD, changed my life and the lives of everyone in my family. We still find it difficult to understand or discuss.*

The pain never goes away for the ones left behind. If you won't do it for yourself then get the help you need for those you say you love.

The Big Picture-A Path Forward

The Mighty Oaks Approach

"How do you do it?"

This question has been asked hundreds of times over the years to just about everyone on the Mighty Oaks Team. The question is then followed up with a statement like:

"I (or my husband, wife, son, daughter, etc.) have been to every program available and not found the level of help experienced here in just a few days-everything is different!"

This is a fair question given the amazing life change that God has allowed us to see as thousands of men and women rediscover the hope and purpose that they thought was gone forever. In the early days when this question was asked, we would reply with something like, "God does the work, we are just thankful to be a part." This was a sincere answer but not typically what the one asking the question was looking for. Really, when someone asks "how?", they are asking about the process. It's one thing to believe that God does the work but another thing entirely to know how to get the most broken people in the world to a place where they will let Him!

Understanding this, we began working to deconstruct our process in a way that made it accessible to everyone. Trauma is a life issue not reserved for Veterans and we wanted to explain what we do in a way that makes sense to those who may never attend one of our programs. This chapter is that explanation. This is not a discussion of the schedule or logistics or even curriculum used in our programs (all of that can be found on our website), but about a clear path for anyone struggling with a loss of purpose, identity, and hope.

Our programs are not specifically designed to end suicide (Veteran or otherwise), but we passionately believe that if everyone lived the process described here that we would indeed eliminate the most pressing epidemic of our time! How can we say this? Because we have lived it. As you have already read in previous chapters, the elimination of suicidal thought and attempts in the lives of those who understand these concepts has been the clearest bi-product of what we teach. Suicidal ideation and, if not addressed, suicide itself, is the fruit of a poisoned tree. Our goal is not to address the fruit, but to get the tree to a point where the fruit becomes healthy.

The Big Picture-A Path Forward

Let's look at that process.

Those who serve in the military all understand (or should) one basic truth of service: The only reason you exist is to fight and win America's battles. Different branches and different jobs all have a place in the one big picture of keeping our fellow citizens safe from the enemies of the world. To that end, billions of dollars are spent each year to answer three basic questions:

1. Who is the enemy?

2. What is the objective: What do we need to do to win?

3. How do we make that happen: What is the vehicle or mode to get us there?

The answers to these three questions shape policy, start production of weapons and vehicles, and drive personnel decisions. These are the questions that must be answered to fight and win. In a very real way, these are also the questions that must be answered if we are going to win the emotional and spiritual battles destroying lives and futures every day.

The answers to these questions form the basis of the Mighty Oaks process.

The Enemy: The Reason we Fight

So often the reason we lose the fight against suicide is because we are looking at the act of suicide as the enemy instead of understanding that self-harm is what happens when the actual enemy is not defeated. This is why, in the Veteran community, in spite of millions of dollars and countless organizations dedicated to solving this problem, or at the very least letting everyone know that there is one, we are still losing! We have decided that simply talking someone out of hurting themselves is enough instead of addressing the REASON they thought they needed to.

So, what is the real enemy? We will get there but first we need to define three words so that we are clear:

> Enemy: An enemy is anyone or anything that gets in the way of accomplishing the mission. In life, our enemies are those things that keep us from living the life we were created to live-the mission given to us by God.

> Weapon: A weapon is a tool used by the enemy to do harm. This is important to understand because sometimes we identify the tools as the enemy. Rifles and explosives are used on the battlefield to do harm, but they are not the enemy. In life, bad decisions, people, and habits can cause

incredible damage. But those are not the enemy. They are tools used by the enemy, often to great effect.

Cause: The cause is the reason something happens. I know this seems intuitive, but I have had many discussions with people who identify their past, what they have done or what has been done to them, as the enemy. While all of those play a part, they may be a cause FOR the enemy, but they are not the enemy themselves. Why do terrorists hate America? I don't know. But they do. There are reasons. On the battlefield though, I must fight the enemy whether I understand the reason (cause) or not. Understanding the cause can be helpful when dealing with an enemy and addressing it may in the future prevent more enemies. Right now, though, on the battlefield, I need to deal with the one trying to do me harm. So it is in life. Cause and Enemy are not the same.

So, what is the actual enemy? What is it that creates all of these problems and, if not addressed, can even lead to suicide? In our experience the enemy that must be defeated has two parts:

1. A lack of purpose

2. A lack of identity

I can hear the objections before you say them. Working with Veterans and Active-Duty military members struggling with Post Traumatic Stress Disorder (PTSD), Combat Trauma Issues, and suicidal ideation, I would imagine we have heard just about every objection. I would just ask, as we do in our programs, how are you doing fighting whatever it is you have identified as the enemy? Maybe the reason it has been so hard to win is because you are fighting the wrong thing. When we correctly identify our enemy, we can start moving toward a place of healing and hope and with those, the many struggles that have us bound begin to fall away.

Here is the phrase we often use when approaching these issues from a different perspective: "If what you are doing isn't working, then why not try something new?"

We have identified the enemy trying to keep us from accomplishing our mission and now we need to understand the objective.

The Objective: The Reason We were Created

Remember, defeating the enemy is not the objective even though the enemy must be defeated if we are going to win! The crazy thing here is that many people DO believe that fighting the enemy is the goal. That's the reason so many fight for so long and never actually win. Since they've never defined the objective, they are fighting a never-ending battle that will only produce exhaustion and hopelessness. The enemy is not the goal but stands in the way of the goal.

Then what's the goal or objective? How do we know if we are doing what we are supposed to be doing?

The Objective: Align your life to the life that God created you to live.

This is a lifelong process that will include many ups and downs, but the goal of aligning your life to God's Purpose is a clear objective that moves us forward and sets us free from the many challenges that keep us in bondage. It is in moving towards this objective that the enemies identified above are defeated. As we grow in our God-given design, we find a purpose that transcends the rest of life and an identity that is bigger than a job, relationship, hurt, or our past. If God is the creator-and we believe He is- then He created the world and us in it with a unique design. We were created to function in a specific way that reflects that design and allows us to grow in our relationship with Him.

The reason we struggle so much in our lives is that, even though we function by going to jobs and doing all of the things we are supposed to do, we are functioning at a limited capacity. We are like high performance race cars hitched to a trailer. We can pull the trailer, but we will never experience the speed and performance the engineer had in mind. When we finally begin living according to our design, we experience the freedom and peace that only comes to those who know why they exist.

When we begin to align our lives to the lives God created us to live, we:

1. Can move forward from a past that has kept us in place. Our past, although we will never forget, loses its grip because we have a new objective.

2. Can transition from one stage of life to another without having the bottom fall out because our purpose and identity is tied to God, who does not change, instead of the stages of life which are ever-changing.

3. Make clear decisions about our life, the people in our life, and the things that we do. "Coping" or "Self-medicating" become unnecessary because we are living according to our creation.

4. Experience healing in relationships since we have a guide that puts those relationships in their right place. We also understand that the world does

not revolve around us and that serving others is a higher calling.

5. Can answer the feelings of hopelessness and despair with the eternal truths given to us by an eternal God.

We must have a clear objective that transcends the ups and downs of life if we are going to live a life that truly makes a difference. It is then, in this life, that we can leave a Legacy that will provide direction for those who come behind us. They follow us as we follow the one who gave us life and we all end up exactly where we were created to be.

The Mode: How We Get There

We have identified the enemy that works to keep us from our objective and have defined the objective that will carry us forward. The last question we need to ask is: "How do we get there?" As important as those are, if we can't deliver these messages to the hurting and help them to begin this journey, then it is all just words. So, how do we get there? I will tell you how we do it.

1. We demonstrate complete transparency. We do not approach any of this from the position of teacher to student but of one broken person who has found hope to another broken person who needs help. In our

programs focused on those in the military we have others who have been in the military tell their story-the dark parts and all- and then explain that they found hope when they understood and started to move toward the objective. Transparency tears down the walls of rejection because of shared, if not identical, experience.

2. We are open from the beginning about what we are going to do and make no apology for it. Even though we are officially a "Faith Based Program," most people who attend our programs are not Christians. They are attending because, in many cases, they just don't have anywhere else to go. We acknowledge that and the fact that we are happy for the opportunity to show them another way, and then clearly tell them what is going to happen. "While you are here, we will present a Biblical blueprint of the life God created you to live." We then ask that they reserve judgment, pay attention, and then contrast their life with the life we are presenting.

3. We explain that it is possible to know God the Creator's design for us by learning from the Bible. We spend time each day teaching

basic principles from scripture that need to be lived out to begin aligning to our created purpose.

4. We spend A Lot of time talking. Since we lead with and personally model transparency, we ask the students to do the same. Most do not open up on day one, but by the end of the week they are sharing thoughts and feelings they have never shared before. They feel comfortable doing so because they have seen this modeled from the leadership team. This time of sharing allows those who have been struggling with thoughts that they could not process to begin processing them in a place and with people who will allow them to process them through the lens of scripture.

5. We talk about the importance of having a relationship with God and what that looks like practically in their lives.

6. We call them to live a life that leaves a legacy worth handing to the generation coming behind. This involves deconstructing the idea that life is about us and communicating the impact that our lives, good or bad, can have on others.

7. We provide a group of people on whom they can lean.

8. We help each student to develop a plan that they can take home and deal with the challenges that brought them to the program.

9. We ask them, once the program is over, to live out the "4 B's"

   a. Be in the Word: spend time every day in the Bible

   b. Be in Prayer: talk to God every day

   c. Be in Fellowship: find a church family that teaches the Bible faithfully and allows you to serve and be served.

   d. Be in contact with those in your life who you can reach out to when you are struggling.

We provide the opportunity for leadership training following the program as well as any support that they may need, from encouragement to counseling, so that they can live out what they have been taught.

We lead with our own stories of God's work as we aligned to His purpose, and then support those who make the decision to live according to His design for them. This may seem like a long process, but I could sum it up with one statement:

> Change happens the minute a person DECIDES to align their life to the life they were created by God to live. We simply work to back them into a corner where they MUST decide.

This can all seem a bit overwhelming, but it is really very simple. "How do you do it?"

1. Identify the enemy: Loss of Purpose and a Lack of Identity

2. Define the Objective: Align to the person you were created by God to be

3. Use the power of transparency to point to the guide (the Bible) and motivate a decision

It's amazing the work that God will do and the hope that can be found when we understand He has a plan for our lives and it is better than anything we could ever do on our own!

What Now

Steps Forward in the Face of Suicide

In this chapter we want to get very practical and specifically address the steps that can be taken in the face of suicide by the three groups of people who need to know:

1. Individuals struggling with thoughts and feelings of suicide

2. Pastoral counselors

3. Friends and family of those dealing with suicidal ideations or attempts

While there are many resources available that deal with suicide prevention and intervention, what is presented here is based on the methodology we have used as a part of our Mighty Oaks trauma programs and the advice we have given to countless pastors and family members. This is not intended to replace the use of suicide intervention care and counseling, but to provide a guide for applying the method of "aligning one's life to the person God created them to be" to the specific issue of suicide and self-harm.

If you or someone you know is in IMMINENT danger of hurting themselves or others, stop what you are doing and call 9-1-1 or the National Suicide Prevention Hotline at 800-273-8255. For all others, here are some steps that can be applied personally or as you seek to help.

For the One Struggling with Suicidal Thoughts and Feelings

1. Pre-decide

The power of the pre-decision is not something that can be overstated. If we allow our emotions to dictate our actions, there is a high likelihood that we will make the wrong decision in a moment of feeling overwhelmed. According to findings reported in the Injury Prevention Journal, one third to four-fifths of all suicide attempts are impulsive. Among people who made near-lethal suicide attempts, 24% took less than 5 minutes between the decision to kill themselves and the actual attempt, and 70% took less than 1 hour. Additionally, 90% of those who survive a suicide attempt do not go on to die by suicide.[13] Many conclusions can be drawn from this data but it is clear that deciding not to hurt oneself before thoughts and feelings of suicide present themselves may be the difference between life and death. Make the decision while you still have the capacity to do so and eliminate suicide as an option.

2. Know Yourself and Seek Self-Improvement

One of the leadership traits that I learned in the Marine Corps was to "know yourself and seek self-improvement." Most of us know, even if we do not like to admit it, where we are weak and where we need to grow as a person. Most of the time these shortcomings do not have a strong impact on our daily lives, but when we are feeling overwhelmed or depressed it is these areas that may cause us to fall. Do not go places or do things that you know have a negative impact on your mental health and grow in areas that have presented problems in the past. For example (and there are many that could be shared), if you get depressed when drinking alcohol, don't drink alcohol. If you find yourself drinking every time you go to a certain place or go out with a certain group of people, stop! Know yourself and grow where you are weak. Don't put yourself in a bad position and your chances of making a bad decision decline dramatically. This applies to both behaviors and relationships. But what if you don't know how?

3. Get Good Counsel

Proverbs 11:14 tells us that there is safety in the multitude of counselors. Find a qualified counselor that can speak truth into your life and provide the perspective needed to align to God's purpose. This should be someone who is both experienced and qualified to apply the principles of Scripture to your life. Often the reason we are overwhelmed and depressed is because our outlook on the future is faulty and our perspective is wrong. Also, there may be an area of sin that needs to be addressed and a counselor who understands how the Bible applies to life can help point that out and show the path forward. Finally, a good counselor can be the key to growth in the area of "Know yourself and seek self-improvement." It is possible to know where you need to grow or what you need to stay away from but not know how. A qualified counselor can provide the direction that you need.

4. Speak to a Doctor

It is possible that you are feeling the way you are because of something physical. Seek qualified medical attention to determine if there is a physical issue that needs to be addressed.

5. Engage in "Whatever Things" Thinking

How often, particularly in the day of 24-hour news and unlimited social media sources, do you go through a day feeling bad but not knowing why? In many ways the prevalence of suicide in our society can be tied to the consistent negativity that is a regular part of our diet. The Bible has the answer for this problem in Philippians 4:8:

Finally, brethren, whatever things are true, whatever things are noble, whatever things are just, whatever things are pure, whatever things are lovely, whatever things are of good report, if there is any virtue and if there is anything praiseworthy--meditate on these things.

Just as in a physical diet, what you consume emotionally and spiritually will either build you up or tear you down. Perhaps the most important thing you can do to change the way you feel is to change the way you think. How do you change the way you think? Feed your mind with only those things that are good and stay away from the "junk food" of negativity being peddled by a broken culture.

6. Observe the 4 "B's"

This was covered in the Mighty Oaks Methodology chapter, but it needs to be included here as a reminder. There are four things you need to consistently do if you want to move beyond feelings of self-harm and destruction:

a. Be in the Word-spend time every day in the Bible.

b. Be in Prayer-talk to God every day.

c. Be in Fellowship-find a church family that teaches the Bible faithfully and allows you to serve and be served.

d. Be in contact with those in your life who you can reach out to when you are struggling.

For the Pastoral Counselor

For those in church ministry, applying the Bible to the various situations of life is a part of the job that can be both fulfilling and challenging. When seeking to help those who are struggling with thoughts or attempts of suicide, the desire to help can often be met with a lack of exactly how to address this potentially life-threatening situation. But we are still called to help. It is imperative that the pastoral counselor (simply one who provides counsel from the position of a pastor) is sensitive to the nature of the issue being presented and is quick to call in help from someone more specifically qualified in this area. Don't be afraid to ask for help.

For the pastoral counselor seeking to be a blessing to those in the church, and perhaps just investigating the true nature of the issues presented, here are some thoughts[14]:

1. Enter the person's world and show him special care amid the stressors tempting him toward suicide. The life-threatening nature of suicidality calls you to amplify your normal attending and relational skills. Don't minimize the impact of your caring, compassionate presence.

2. If the person hasn't mentioned suicide but you suspect it, ask. Broach the subject. Reject the myth that raising it might plant it in his head and thereby encourage it. The person who manifests signs has likely considered suicide at some point. Raising the topic provides a safe place to talk about it. The severity of this issue trumps any potential risk.

3. Work hard to understand both his situational pressures and his internal beliefs and motives. Suicidal people operate with an inner logic, however distorted it might

seem to you. While it generally involves some form of "life is not worth living," or "I'm not worth living, so I will end my life," there is no single cause for suicide (e.g., not every suicidal person is depressed). Get to know this person.

4. Minister the gospel. Bring the hope of Jesus Christ and His multi-faceted saving provisions. This includes caringly and carefully unearthing his heart responses to his pressures and presenting the life-transforming beauty and power of Christ.

5. Plead personally with the person not to commit suicide. Appeal to him. Use rational and relational reasons to persuade him.

6. Remember that the person whose life is so miserable that he thinks he needs to end it probably does need to end that life and find in Jesus a new life. Consider Jay Adams's concept of "sympathetic disagreement" in which you sympathetically agree with the person's diagnosis ("my friend, this life that you're living does need to end") but sympathetically disagree with his contemplated remedy ("but I have a way to end that life that is a million times better than your way").

7. Don't minister alone. As much as the counselee allows, seek to connect him to a few mature, caring believers, especially those he admires or respects (maybe within his small group). Work with them as a care team to help watch, care for, and encourage the person.

8. Take the person's threats seriously; err on the side of safety. Don't assume threats or attempts are merely cries for help—they might be threats or attempts to kill himself. Don't ignore those cries.

9. Remember this bottom line: suicide is the person's choice. Apart from institutionalization, you can't stop a person from finding a way to kill himself. But you are not responsible for his choice. As a caring counselor, you are particularly prone to this lie. Don't own another person's wrong decision.

While you can't solve all the problems a suicidal person faces, you can pleadingly and prayerfully point him to the One who can, and you can commit to walking with him through his struggles. Pray that he will gain a sufficient-for-now sight of Jesus that will make him choose what Simon Peter chose when others turned away from Jesus: "Lord, to whom shall we go? You have the words of eternal life. We have come to believe and to know that you are the Holy One of God" (John 6:68-69).

For Friends and Family

There are few things in life more difficult than watching someone you care about struggle. When they are struggling with thoughts of ending their life, the pain of watching that struggle can be overwhelming. The need to do SOMETHING is often motivated by the question, "What might happen if I do nothing?" There is no perfect answer since every situation is different, but in addition to everything mentioned above, here are some thoughts for family and friends:

1. Remember that personal transparency leads the way. Often, when those who you care about are unwilling to talk about what they are going through it is helpful to share a struggle that you have had personally. You enter their world by being honest about your own shortcomings. Don't compare hurts-just express that everyone, including you, does.

2. Always offer Hope. Not false hope, but the hope that with the right help and enough time there is nothing that can't be overcome.

3. Listen. Let them talk and feel. This can be extremely difficult if you want to "fix" the problem but often what is needed most, at least right now, the need to just talk.

4. Be a friend. Do what you would want someone else to do for you.

5. Realize that you do not have to have all of the answers. Seek outside help.

6. Be prepared for pushback but don't take it personally. This is a part of the process.

7. Leave the door open. You never know when a person will be ready for help. Don't shut doors because you are angry or frustrated with a lack of progress.

8. Find others who have experienced helping those they love through similar situations. While offering hope it is easy to lose hope. Find those who have succeeded and have your own hope renewed.

This lengthy chapter is intended to give you some tools you may not have had as you work to defeat the threat of suicide. There is much more help available but perhaps this can serve as a starting point whether you are the one struggling, or you are doing your best to help those for whom you care. Whether you are the one in need of help or the one offering it, refuse to let fear keep you from exposing your own need and hang on to the truth that God has a plan for every life and that it is good!

Hope

Over the course of my life I have come to realize that most of us, even though we do a pretty good job of hiding it, feel like we are standing on the precipice just waiting for someone to push us off! Hopefully this is not all the time, but if the truth were known, we all feel this way from time to time. Life-and all of its problems-becomes overwhelming and exhausting and we are not sure that we can push forward much longer.

One of the reasons so many people like the Psalms is because they often express what we feel and give us hope when we can't seem to find it on our own.

When we are exhausted, we are told that we can rest in hope:

Psalm 16:9 Therefore my heart is glad, and my glory rejoiceth: my flesh also shall rest in hope.

When we aren't sure we have the strength to move on we are given the hope that God will provide what we need.

Psalm 31:24 Be of good courage, and he shall strengthen your heart, all ye that hope in the LORD.

While we are thankful for these and other passages from scripture that give hope, we still struggle. Why is it so hard? I think that most of us struggle in this area of Hope because we lack an accurate definition. We conclude that hope is just some unknowable thing that only those detached from reality can have. If we cannot understand hope, how can we have it? Here is a definition that has really helped me:

Hope: The confidence that where I stand right now is not all there is!

We all live inside of a very small circle and most of the time our reality is what is happening right now inside of that circle. Hope tells a different story. Hope acknowledges what is happening in the circle of our reality, but it raises our view to something bigger that transcends what we are living through. Hope says that there is more to this life than what we can see, and even though we may not have control, there is one who does. Hope serves as an anchor when we feel lost and tossed in the storm.

Hebrews 6:19 Which hope we have as an anchor of the soul, both sure and steadfast, and which entereth into that within the veil;

I love this truth because it tells us that our hope is found IN the storm not after it. We are not waiting from deliverance to find hope, we can have it now when we need it most!

> *Romans 5:1 Therefore being justified by faith, we have peace with God through our Lord Jesus Christ: 2 By whom also we have access by faith into this grace wherein we stand, and rejoice in hope of the glory of God. 3 And not only so, but we glory in tribulations also: knowing that tribulation worketh patience; 4 And patience, experience; and experience, hope: 5 And hope maketh not ashamed; because the love of God is shed abroad in our hearts by the Holy Ghost which is given unto us.*

Where do we find this hope?

> *1Timothy 1:1 Paul, an apostle of Jesus Christ by the commandment of God our Savior, and Lord Jesus Christ, which is our hope;*

Jesus is our hope because, on the cross, He made forgiveness, new life, and life eternal possible for all who believe!

1. Paid for sin

Romans 6:23 For the wages of sin is death; but the gift of God is eternal life through Jesus Christ our Lord.

1Peter 2:24 Who his own self bares our sins in his own body on the tree, that we, being dead to sins, should live unto righteousness: by whose stripes ye were healed.

2. Defeated death

2Timothy 1:9-10 Who hath saved us, and called us with a holy calling, not according to our works, but according to his own purpose and grace, which was given us in Christ Jesus before the world began, But is now made manifest by the appearing of our Saviour Jesus Christ, who hath abolished death, and hath brought life and immortality to light through the gospel:

3. Gives new life

2 Corinthians 5:17 Therefore if any man be in Christ, he is a new creature: old things are passed away; behold, all things are become new.

4. Secured eternal victory

1Corinthians 15:57 But thanks be to God, which giveth us the victory through our Lord Jesus Christ.

Hope is found when we finally stop viewing the world from our own position and instead begin to view the world from our position in Christ! It is in having a relationship with Christ that provides real hope and it is that hope that carries us forward in a world that often seems hopeless.

If you have not already will you decide today to stop trying to figure it all out and instead call out to the one who already has?

One more verse, Romans 10:13:

> *For whosoever shall call upon the name of the Lord shall be saved.*

Saved from what? From sin, death, hopelessness and an eternity separated from the God who created us. Call out to Him today and experience the hope that

He alone can provide.

MIGHTY OAKS
FOUNDATION

Mighty Oaks Foundation is a faith-based veteran service organization offering nonclinical, peer-based mentoring to our nation's Warriors. Our resiliency and recovery programs, open to Veterans, First Responders, and the families of both, are intensive peer-based programs offering instructional sessions rooted in biblical principles and taught by those same Warriors who were once struggling and have since renewed their purpose through faith. During our resiliency and recovery programs, we focus on camaraderie building through various team building activities designed to challenge our Warriors to overcome their past experiences to move forward with renewed purpose. Spiritual battles are not an American issue, and we realize that. America's Warriors remain at the forefront of the Mighty Oaks mission as we continue to serve tens of thousands per year, but the Lord has called us internationally as well to serve our allied military partners throughout the world. For that reason, you will find Mighty Oaks personnel all around the world, even in the most austere environments, as we continue to fight for the one true kingdom: the Kingdom of God. Join us as we **SAVE LIVES**, **RESTORE FAMILIES**, and **CHANGE LEGACIES**.

(★) **RESILIENCY PROGRAMS**　　　**INTERNATIONAL PROGRAMS** (★)
(★) **RECOVERY PROGRAMS**　　　**AFTERCARE PROGRAMS** (★)

"To restore the brokenhearted through Christ, to build leaders of leaders to rise up from the ashes; they will be called Mighty Oaks of Righteousness." — Isaiah 61

MightyOaksPrograms.org

LEARN MORE

This book was made possible through
the successful partnership with:

The Warrior's Journey

The military creates a set of unique challenges that most people do not understand These invisible wounds too often lead to isolation, addiction, and even suicide. The Warrior's Journey (TWJ) provides warrior-to-warrior intervention and preventative services and resources to help heal these invisible wounds. TWJ is a community of warriors helping warriors find true resilience. No paperwork. No red tape. No bull.

Prevention

TWJ raises awareness and offers preventative services and resources, at no cost, to help engage, educate, support, and prepare our warriors for their time in the military and beyond as they transition back to civilian life.

Intervention

TWJ has built a team of veterans to provide real-time crisis intervention. Our individualized and relational approach to intervention is always confidential and focused on navigating the available help and resources provided TWJ and their 72 partner organizations, to include Mighty Oaks Foundation.

To learn more about The Warrior's Journey, scan the QR code below or visit *twj.org*

Endnotes

1 "Suicide Among Veterans and Other Americans 2001–2014" (PDF). Mentalhealth.va.gov. Retrieved 1 June 2019.

2 "DoD Releases Report on Suicide Among Troops, Military Family Members. Defense.gov September 26, 2019

3 PREVENTS, The Presidents Roadmap to Empower Veterans and End a National Tragedy of Suicide, June 2020, U.S. Department of Veterans Affairs

4 Suicide and Suicidal Attempts in the United States: Costs and Policy Implications, https://www.ncbi.nlm.nih.gov/pmc/articles/PMC5061092/

5 Children Who Lose a Parent to Suicide More Likely to Die the Same Way, April 2010, https://www.hopkinsmedicine.org/news/media/releases/children_who_lose_a_parent_to_suicide_more_likely_to_die_the_same_way

6 Pitman AL, Osborn DP, Rantell K, King MB. Bereavement by suicide as a risk factor for suicide attempt: a cross-sectional national UK-wide study of 3432 young bereaved adults. BMJ Open. 2016;6(1):e009948.

7 Ibid.

8 Ibid.

9 Sveen CA, Walby FA. Suicide survivors' mental health and grief reactions: a systematic review of controlled studies. Suicide Life Threat Behav. 2008;38(1):13-29.

10 Kučukalić S, Kučukalić A. Stigma and suicide. Psychiatr Danub. 2017;29(Suppl 5):895-899.

11 Spillane A, Matvienko-Sikar K, Larkin C, Corcoran P, Arensman E. What are the physical and psychological health effects of suicide bereavement on family members? An observational and interview mixed-methods study in Ireland. BMJ Open. 2018;8(1):e019472.

12 Ibid.

13 Sonja A Swanson, Mara Eyllon, Yi-Han Sheu, Matthew Miller. (2020) Firearm access and adolescent suicide risk: toward a clearer understanding of effect size. Injury Prevention 160, injuryprev-2019-043605.

14 Portions of this text used by permission from Robert Jones, Nine Guidelines for Counseling Suicidal People, Biblical Counseling Coalition, https://www.biblicalcounselingcoalition.org/2019/09/09/nine-guidelines-for-counseling-suicidal-people/